What You Won't Do
-FOR-
Love

T0021005

A CONVERSATION

DAVID SUZUKI
TARA CULLIS
MIRIAM FERNANDES
RAVI JAIN

COACH HOUSE BOOKS, TORONTO

first edition

Canadä

Published with the generous assistance of the Canada Council for the Arts and the Ontario Arts Council. Coach House Books also acknowledges the support of the Government of Canada through the Canada Book Fund and the Government of Ontario through the Ontario Book Publishing Tax Credit.

LIBRARY AND ARCHIVES CANADA CATALOGUING IN PUBLICATION

Title: What you won't do for love : a conversation / David Suzuki, Tara Cullis, Miriam Fernandes, Ravi Jain.
Other titles: What you will not do for love
Names: Suzuki, David, author. | Cullis, Tara, author. | Fernandes, Miriam, author. | Jain, Ravi, author.
Identifiers: Canadiana (print) 20220204594 | Canadiana (ebook) 20220207291 | ISBN 9781552454541 (softcover) | ISBN 9781770567283 (EPUB) | ISBN 9781770567313 (PDF)
Subjects: LCSH: Suzuki, David, 1936-—Drama. | LCSH: Cullis, Tara—Drama. | LCSH: Fernandes, Miriam—Drama. | LCSH: Alvsvaag, Sturla—Drama. | LCGFT: Drama. | LCGFT: Creative nonfiction. | LCGFT: Biographical drama.
Classification: LCC PS8637.U97 W43 2022 | DDC C812/.6—dc23

What You Won't Do For Love is available as an ebook: ISBN 978 1 77056 728 3 (EPUB), 978 1 77056 731 3 (PDF)

FOREWORD

by David Suzuki

How can we convey an important issue in a way that people understand what it is, how it affects all of us, and what can be done about it? That is the challenge that I have grappled with for decades.

I was astounded at the speed with which scientists identified and isolated the COVID-19 virus, then sequenced the viral RNA and determined its mode of attack through spike proteins. Out of dozens of ways to try to stop the virus, a revolutionary method – using mRNA of the gene to specify a spike protein rather than the spike protein itself – has proved extremely effective at eliciting an immune response.

In the twenty-four-hour news cycle that rapidly burns through stories, media have been saturated with COVID-19 information, not for days or weeks but for over two years now. It has been an emergency, and media have treated it as such. In contrast, climate change was recognized in 1988 as a threat to human survival 'second only to a global nuclear war' at an international conference on the atmosphere in Toronto, which called for a 20 percent reduction in greenhouse gas emissions over fifteen years. That same year, the IPCC (Intergovernmental Panel on Climate Change) was established to be the most authoritative source of information. If nations had taken the target seriously, we could have completely avoided the climate crisis we are in today. Media failed to cover climate change with the same endurance and intensity as they did with COVID-19.

Year after year, scientific reports on climate change have taken on an ever-more-urgent tone, while the evidence of its impact – through massive forest fires, floods, drought, storms, and melting glaciers and icefields – has dramatically increased. But until recently, these weather events have been reported independently, as if they were not necessarily connected by the driving force of climate change. In 2018, a special IPCC report urged the world to avoid catastrophic temperature rise of more than 1.5 °C above pre-industrial levels by reducing emissions by 45 percent by 2030 and by 100 percent by 2050. The day after the report was released, marijuana became legal in Canada and wiped away any further reports about the IPCC targets.

Nature is the source of the air we breathe, the water we imbibe, the soil that generates our food, and the energy in sunlight that fuels our bodies. And nature remains the only way that carbon can be removed from the atmosphere and sequestered. In 2019, the United Nations reported that the planet is undergoing a human-caused mass extinction of species, the repercussions of which are devastating in countless ways. Yet the day after the report was released, Harry and Meghan had a baby and obliterated further discussion of species extinction and climate change in the media. What will it take to evoke the kind of sustained emergency response and discussion that we saw over the COVID-19 pandemic?

That's the question I have been pondering since 1962, when, after living in the U.S. for eight years to get my post-secondary education, I returned home to Canada to begin a scientific career in genetics. But that year, Rachel Carson published *Silent Spring*, which documents the impact of pesticides on the natural world, and it hit me like a thunderbolt. So I was swept up with millions around the world in what became the modern environmental movement. Reading the book, I realized that scientists focus on a single part of nature, isolate that fragment, control everything impinging on it, measure everything within it, and in that way acquire powerful insights. But in the wider world, everything is connected in ways we barely understand. As powerful as science is, the very process of focusing on individual elements is the essence of reductionism, and it blinds researchers from the context within which that isolated piece of nature exists and interacts.

A flask, growth chamber, or even field plot fails to address that interconnectivity; when DDT was sprayed on farmers' fields, rain, wind, and snow swept the insecticide into the air where it spread and flowed into rivers and lakes, and its concentration increased up the food chain until, in the shell glands of birds and the breasts of women, the pesticide was concentrated tens of thousands of times. Scientists only discovered 'biomagnification' as a biological phenomenon by tracking down the cause of sudden huge declines in raptors like bald eagles.

When I arrived on campus, the University of Alberta was producing a television program called Your University Speaks, broadcast by the local CBC station on Sunday mornings. I was asked to do an exposition on genetics, and the producer liked it so much I ended up doing a series of eight shows. After a couple of programs had been broadcast, I ran into people who would tell me they had seen and liked the show. I was flabbergasted to find that anyone would watch television at that time of day and on Sunday, and that's when I realized that television had become a powerful medium of communication.

Popular media like the press have whole sections on politics, finance, celebrity, and sports, but they ignore the reality that the most powerful force shaping our lives and society is not politicians, the economy, athletes, or famous people, it is science as it is applied by industry, medicine, and the military. Consider that, when I was a child, my parents wouldn't let me go to movies or public swimming pools in summer because they feared I might catch polio. Back then, millions died of smallpox, which is now extinct; there was no television, computers, cellphones, satellites, plastics, organ transplants, or antibiotics. Scientists didn't know how many chromosomes humans have, what DNA did, or how sex was determined. Science has transformed our lives and society since then.

I deliberately chose to popularize science as well as do research because I believe the general public has a huge stake in the consequences of science and technology and therefore should be literate enough to have an opinion on how these applications should be made or restricted, rather than allowing businesses, politicians, and military officers alone to determine the fate of new technologies. So I chose to use television and radio to explain science whenever the opportunity arose.

My hopes of elevating scientific literacy through television have been dashed by cable and the internet. With so much information available, most people choose to surf television and the internet to find people, sites, and programs that confirm what they already believe so they don't have to consider the facts or change their minds. If they want to believe climate change is a hoax, that vaccines give others the means to control them, that Earth was created in six days by God, or that the planet is flat, there are numerous websites, often with 'experts' with PhDs, devoted to validating these beliefs.

I have written books (many for children) and weekly columns, and I have appeared on radio, television, and podcasts. It is clear that reason and facts alone no longer suffice to move people and society to action. We have to touch people's hearts, to move people emotionally. That's why the arts, especially music, have an important role to play in the environmental movement. But most of my communication in the popular media has been one-way, not a dialogue or an in-person exchange of ideas. So when Ravi approached me with the idea of a play, I did not immediately dismiss it because I was intrigued by the possibility of a more intimate personal conversation with an audience, a conversation that might convey passion, fear, and love, emotions that are as important as facts alone. You can judge from this play whether I am right.

CREDITS

Although originally intended as a theatrical production, because of COVID-19 *What You Won't Do For Love* was first produced as a film in fall 2021 and then received its premiere as a theatre production in summer 2022. This book features text from the live production alongside images from the film.

In the theatre production, MIRIAM, DAVID, TARA, and STURLA sit around a dinner table. Behind the table there is a large projection screen. During the interlude segments, the audience is immersed in images and sounds of nature, music, and poetry.

2022 THEATRICAL PRODUCTION

World Premiere run, June 9–19, 2022

Luminato Presents
What You Won't Do For Love
A Why Not Theatre Production

Originally co-presented and produced with support from TO Live and Soulpepper
Written by Tara Cullis, Miriam Fernandes, Ravi Jain, and David Suzuki, with additional text by Sturla Alvsvaag
Performed by Tara Cullis, Miriam Fernandes, Sturla Alvsvaag, and David Suzuki
Directed by Ravi Jain
Associate Directed by Miriam Fernandes and Kevin Matthew Wong
Music and Sound Design by Meg Roe
Projection Design by Jamie Nesbitt
Lighting Design by André du Toit
Production Managed by Joshua Hind
Produced by Kevin Matthew Wong
Dramaturgy by Broadleaf Theatre

2021 FILM

Released December 3, 2021

What You Won't Do For Love
A Why Not Theatre Film
Co-presented and produced with support from TO Live and Soulpepper

Written by Tara Cullis, Miriam Fernandes, Ravi Jain, and David Suzuki
Performed by Tara Cullis, Miriam Fernandes, Sturla Alvsvaag, and David Suzuki
Directed by Ravi Jain and Kevin Matthew Wong
Produced by Kevin Matthew Wong
Executive Produced by Ravi Jain, Kelly Read, Miriam Fernandes
Music by Meg Roe
Original Performance Conceived by Ravi Jain
Film Production by Pool Service Productions

FOR WHY NOT THEATRE:

Co-Artistic Director and Founder – Ravi Jain
Co-Artistic Director – Miriam Fernandes
Executive Director – Kelly Read
Finance and Administration Manager – Kira Allen
Administrator and Office Manager – Katy Mountain
Partnerships Manager – Sam Semczyszyn
Marketing and Communications Manager – Yvonne Lu Trinh
Marketing Coordinator – Gabriella Albino
Production and Technical Manager – Crystal Lee
Managing Producer, MAKE Platform – Kevin Matthew Wong
Managing Producer, SHARE Platform – Michelle Yagi
Managing Producer, PROVOKE Platform – Tom Arthur Davis

SCENE 1
INTRODUCTION

MIRIAM: Hi! My name is Miriam. I'm a theatre artist from Toronto.

A few years ago, I, like most of you, was becoming increasingly concerned about the state of our planet, and what I was doing to it. And so, as an artist, I thought: I'm a storyteller – I can tell a story to convince people that we need to change our behaviour, and maybe we can change the world. And I thought: who better to tell that story with than David Suzuki?

So I reached out to David with an email inviting him to play the title role in a play called *The Life of Galileo.* David says –

DAVID: Dear Miriam,

I am blown away by your offer. Wow, what an incredible honour. I am not an actor and have never taken such a role. Quite honestly, if I was thirty years younger, I would have leapt at the chance, but now the thought of memorizing long segments is pretty daunting. But before I plead senility, is it possible to see the script to see how challenging it might be? I would love to have a chance to consider it.

Sincerely, David

MIRIAM: Wow! That's amazing! David Freaking Suzuki wants to do a play with me. He says:

DAVID: As long as I don't have to memorize anything, I'm in.

MIRIAM: So I promise him: no memorizing lines. We'll just read everything right from the page.

DAVID: Really? Are you sure?

MIRIAM: Yes, you can do anything in the theatre!

So, we email a few ideas back and forth, and after about a year we meet in Toronto for a coffee and he says:

DAVID: So, how long is this going to take?

MIRIAM: Plays can take a while to make, maybe two or three years?

DAVID: Well, you know, you'd better hurry up with this idea of yours, I'm eighty-two and I'm going to die soon.

MIRIAM: I'm taken aback. I imagine what the world without David Suzuki would be … I could tell he was only partially kidding, but I just … I um and ahhh … and I had noticed he talked a lot about his wife, Tara, so I say, 'I know you are really busy, so what if we make this quality time with your family and we can do this play with you and Tara?'

He laughed and said:

DAVID: Well, if you can convince her, be my guest …

TARA: So she convinced me!

MIRIAM: Yes! Tara's in! Now, at that same time, my fiance, Sturla, and I were figuring out our future. We'd just finished grad school and were at the end of our twenties, and it felt like all these decisions needed to be made.

STURLA: Like where were we going to work, when we were going to get married, if we were going to have children, start a family …

MIRIAM: The thought of bringing kids into this world terrified me.

STURLA: But the first big question was: where are we going to live? I was living in Norway, Miriam was in Canada.

MIRIAM: And we had spent three and a half years flying back and forth, which was not very sustainable for the environment –

STURLA: Or our relationship.

MIRIAM: So we're in the middle of this mess, I'm stressed about the future – both my own future and the planet's – and I'm stressed about trying to make this play with David and Tara, when Tara says:

TARA: Sturla, you're an actor, right?

STURLA: I am.

TARA: Miriam, why doesn't he just join us?

MIRIAM: Huh. Yeah, right. So he does, and in January 2019 we all meet in Vancouver to write a play. We start with conversations – interviews – and sifting through their photos and their stories of all of their adventures.

I expected us to talk about science, but the whole time I couldn't stop thinking about love ... These two have spent their entire lives fighting what feel like impossible battles – how did they do it? Maybe it's love? Spending time with them, their love for each other, their family, the planet ... it's infectious ...

What if we could love the planet the way that they love each other? Would we change?

STURLA: December 2019, I officially immigrate to Canada.

DAVID: January 2020, we meet again in Vancouver to record our conversations –

TARA: And we make a play out of those conversations.

STURLA: February 2020, we perform in front of a live audience and they love it –

TARA: Yeah they did, it was so fun – I'd never performed onstage before ...

DAVID: We were all set to take our show across Canada!

MIRIAM: And then March 2020.

COVID shutdown. Income inequity, racial injustice, climate emergency – fires in Australia, California, Oregon, and Canada ... is this the kind of world I want to bring a child into? In the isolation and the sadness, the only solace I found was in reading those conversations we'd had with David and Tara. And every day our play felt more and more urgent.

TARA: And we couldn't do our play because the theatres were closed!

DAVID: So September 2021, we adapted ...

MIRIAM: We turned our play into a film.

TARA: Oh, it was so much fun! I'd never been in a film before!

DAVID: Miriam, for the film, I didn't have to memorize any lines – what about this play?

MIRIAM: David, we're going to read right off the script – just like right now.

DAVID: Really? We can do that?

MIRIAM: I told you we can do anything in the theatre!

STURLA: They seem to be enjoying it so far.

TARA: Oh yeah! It's so exciting! Y'know – I've never performed in a festival before!

STURLA: You're doing great!

TARA: Am I? Oh!

DAVID: Okay. Now I think that we should start this play with a joke –

TARA: I thought we were starting with the poem?

MIRIAM: Don't worry, Tara, we are ...

DAVID: Or what if we start by establishing our hypothesis?

STURLA: I think we're going for something more poetic ...

TARA: I love that poem.

STURLA: Yeah – it's beautiful.

DAVID: But if we can establish a hypothesis, we can measure the impact and results of our play.

STURLA: David, this is a story, not a science experiment. We want to touch people's hearts.

MIRIAM: Yeah, we're storytellers, we're not scientists.

DAVID: Actually, I *am a scientist.*

MIRIAM: Fine. We're not doctors.

STURLA: Miriam, they're both doctors.

TARA: We're not medical doctors. We both have doctorates.

MIRIAM: Right, okay fine! Can we all agree that we all want to find a way to communicate the science?

EVERYONE: Yes.

MIRIAM: To tell this story better?

EVERYONE: Yes.

MIRIAM: Great. So here we are. In a theatre. And let's imagine our story begins with us at your cottage.

TARA: But I'm just wearing my clothes. What about a costume?

STURLA: This is your costume.

TARA: I always wanted to wear one of those Elizabethan dresses –

MIRIAM: Let's save it for the Stratford run –

TARA/DAVID: WOW! We're going to Stratford?!

MIRIAM: NO! NO! Focus. Here we are, spending a day together, and these are our conversations. And somewhere in here, in this play, there are patterns. There is a message. It's a feeling. It's poetic. Tara?

TARA: Okay. Ahem.

INTERLUDE
SONNET 18

Shall I compare thee to a summer's day?
Thou art more lovely and more temperate:
Rough winds do shake the darling buds of May,
And summer's lease hath all too short a date;
Sometime too hot the eye of heaven shines,
And often is his gold complexion dimm'd;
And every fair from fair sometime declines,
By chance or nature's changing course untrimm'd;

But thy eternal summer shall not fade,
Nor lose possession of that fair thou ow'st;
Nor shall death brag thou wander'st in his shade,
When in eternal lines to time thou grow'st:
    So long as men can breathe or eyes can see,
    So long lives this, and this gives life to thee.

William Shakespeare

DAVE SUZUKI

SCENE 2
TARA

DAVID: Back in the 1930s, English people with higher education would come to Canada and get good jobs: many of them were in the civil service and were in charge of rounding up Japanese Canadians and shipping us off to internment camps.

TARA: This is during World War II – they confiscated all their houses and possessions, everything ...

DAVID: Yeah, so my dad really blamed the English for our incarceration.

So, years later when I was in high school, Dad realized, I was always ogling girls, right? But it would never have occurred to me to ask a white girl out. I was too afraid of rejection. Anyway, Dad called me in and he said, 'David. Listen, I know that you wanna start going out with girls. I want you to understand that the only acceptable mate for you in this family is a Japanese girl.'

'Dad, there are only ten Japanese girls in London and three of them are my sisters.'

He says, 'Okay okay, yeah, a Chinese girl would be okay.'

I say, 'Dad there's only three Chinese families in London, I don't know any of them – '

'Okay okay okay, if you find a Native girl, that's okay – '

'I've never even met a Native girl, I – I don't know where the reserve is – '

'Okay. A Black girl would be all right.'

'Uh, the only Black girl I know is Annabelle Johnson, and I know she's not interested in me – '

'Okay then, a Jewish girl.'

So it was all based on race and discrimination, and being aware of being discriminated against. And at the bottom of the list for my dad was, 'An English girl, absolutely if you have anything to do with one, I will kick you out of the family.'

An English girl. No way.

TARA: So I was born in England.

MIRIAM: Of course you were.

TARA: I emigrated to Canada when I was six, grew up here and, much to David's father's surprise, three weeks after David and I met, we were engaged.

MIRIAM: Three weeks! ... It took us three years!

TARA: I wasn't in a rush – but he was!

STURLA: Did your dad kick you out of the family?

DAVID: No! You know, it took a while – well, that's another story.

TARA: Yeah, you made him cry.

DAVID: Yeah, it was the only time I'd ever seen my father cry.

STURLA: What happened?

DAVID: Well, we were at my parents' house and Dad started in on Tara, saying things like 'Your people did this, and your people did that,' and then he said, 'The English were eating with their bare hands when the Japanese were eating on fine china with ivory chopsticks.' So Tara was really upset.

TARA: I wasn't that upset. I mean, it was true!

DAVID: Well, I was upset, and I dropped Tara off and went back to my dad and I yelled at him and said, 'You're nothing but a goddamn bigot. Just because you experienced racism doesn't give you the right to become a bigot yourself. When you become a bigot then the racists win.' And, you know, that shook him to have his son call him a bigot. And after some time I think he changed the way he looked at the world, and then Dad *really* fell in love with Tara. How could he not?

TARA: What did your parents think about you marrying a Norwegian?

DAVID: Didn't they want you to marry an Indian guy?

MIRIAM: No, my sister covered that base. My parents really love Sturla, so that wasn't our problem.

STURLA: Our problem was that we lived in different countries!

MIRIAM: Yeah, so much of our relationship was long distance between Norway and Canada.

TARA: Oh gosh, how did you do it? We had two years of long distance and just spent all our time crying on the phone.

STURLA: Yeah, we had many tearful goodbyes at airports.

MIRIAM: All that time apart was spent saving up our money and figuring out when we could see each other next.

STURLA: But there's nothing like that feeling of seeing each other again at the airport after three months away and an eight-hour flight. Seeing her there it was like no one else existed.

DAVID: Hey, do you remember when I visited you in Ottawa, and we were in the restaurant?

TARA: Oh, ha, the guy who interrupted our 'romantic dinner'?

DAVID: So this is the first time I'm back in Ottawa after Tara and I met. It was freezing cold and I was absolutely in the throes of love! It was in February and we went to this restaurant, and it's very dark, and we're having this romantic dinner –

TARA: Our food had just arrived –

DAVID: And this guy bursts into the restaurant, and he's yelling, 'I don't know where the fuck I am! What's going on?'

And the owner gets on the phone to 911, and this guy is reeling around the restaurant, banging on tables, and everyone, including me, is kind of hunkered down.

TARA: And he ran out into the traffic –

DAVID: And he's totally crazed –

TARA: Bumping into cars – I thought he was going to get run over! So I said, 'David we've got to do something, he's going to get hurt.'

DAVID: I'm in love with her. I'm just glad he's gone so we can go back to having our romantic dinner. But I can't be an asshole if she asks me to do something, so we get up.

TARA: Put our coats on, out we go, and so we went up to him on the corner –

DAVID: No, he was in the middle of the street, cars honking by, and we pulled him off –

TARA: Okay, on to the corner and he's still saying, 'I don't know where the fuck I am!' and I say, 'Well, you're on Bank Street – '

DAVID: And then suddenly he hauled off and slugged her in the face.

TARA: He hit me. Yeah!

DAVID: And I grabbed him – I was so mad – I grabbed him and he just shrank – he was wearing this big puffy jacket, he seemed really big. When I grabbed him, he just shrank into this little wee guy. And I pulled him into the restaurant and I said, 'Don't you do that again!'

MIRIAM: (*To Tara*) I would've been too scared to get involved.

TARA: Well, I was scared! I was scared the guy would get hurt. He was in danger, someone's got to do something –

And usually if one person jumps up and does something, then everyone else wants to help as well. But they all have their head down for a while, hoping it doesn't have to be them. I guess I really think about what would happen if I didn't do anything and he was hit by a car and a little puddle of squished blood says to you, *you just sat there*? But I had not expected to get punched in the face.

MIRIAM: Huh, you know, that reminds me of this time when we were in New York a few years ago. We were on the subway and this guy was in between the cars, threatening to jump off the moving train – and nobody was doing anything –

STURLA: Everyone was just looking at their phones and ignoring him, like it was an everyday event –

DAVID/TARA: Like in that restaurant.

MIRIAM: So, I was frozen, terrified, and Sturla just goes over to him and very calmly just talks to him, inviting him to come back into the train – and finally he gets off the ledge and comes back into the train ...

STURLA: And then when he gets off at the next stop, all these strangers started coming up to us saying *thank you* and *good job* and all – and I kept thinking how strange it was that everyone just ignored him. It was like people had gotten used to not doing anything, thinking that their actions don't matter, so even if they did do something, the same thing would just happen again tomorrow. But, like you, Tara, I was worried that something horrible would happen to him.

TARA: But then you realize everyone wants to do the same thing, they just need to know they are not alone. But yeah, it's scary to do something like that. You can feel exposed.

DAVID: Ever since I've known her, Tara has always been like that. She'll just roll up her sleeves and get it done. That's how she started the foundation!

MIRIAM: Tara, you started the David Suzuki Foundation?

TARA: He says I dragged him into it!

DAVID: It was your idea!

TARA: Yeah, he was going around telling everyone bad news all the time, getting everyone depressed, and I said to him, 'David, you can't just tell people the problems, you've got to give them solutions! People need answers, action, hope!'

MIRIAM: This is in the nineties?

TARA: Well, we were just coming out of the eighties, which had been such an electric time for us. We had our kids while I was getting my PhD, which was dificult, and my first job was teaching at Harvard –

DAVID: I was so proud of her!

TARA: I really loved my job. I was teaching writing about literature, and about social and ethical issues –

DAVID: She was teaching literature students, and eventually she asked to teach the science students –

TARA: Yeah, I knew how important science is in society –

DAVID: You'd better, you were married to me!

TARA: I realized, my god, we need articulate scientists, we need scientists who can write. They need to be able to tell a good story. And that was really something, because my literature students wanted to learn how to write, but the science students – these Westinghouse Award winners – did not want to learn how to write.

STURLA: Why is that?

TARA: Well, I think the scientists were afraid that if they wrote plainly for the general public that they would be revealed as not very smart!

DAVID: A lot of people are like that: it's not just scientists.

TARA: That's why people use a long word when a short one will do. The other day I used 'preternaturally' when I could have just said 'very.' Anyway, there I am teaching at Harvard, but in the summers I'd be up to my elbows in these environmental projects – you know, fighting to prevent the logging in the Stein Valley, and dams at Site C, and Altamira in Brazil. So for five years I was living between these two worlds, and it was hard. But then one day I had this realization: my students had every advantage in life, they didn't need me, they were going to be just fine without me, but the trees and valleys and the Amazon, if we don't fight for them now, they'll be gone.

So I quit.

DAVID: I'm an academic – it broke my heart –

TARA: I wanted to put all my effort into stopping the destruction and the logging and the damming we were protesting at places like South Moresby. We were part of a community that was having an impact, and there was so much energy and collective desire to *do* something *more.* So we gathered a bunch of friends and activists and had our first meeting, and the idea of a new organization emerged.

DAVID: See, our group wanted to start something that would focus on the underlying root causes of the environmental crisis. We didn't want to just put out brush fires, we wanted to extinguish the underground source of the fire.

TARA: And everyone said, 'Oh yes yes, we've got to do this.' And nothing happened. A year later, we had a second meeting and they said, 'Oh yes yes, this is important and we want to do something. We *have* to do something!' And nothing happened!

I realized, no one's ever going to do anything unless I do it. And I thought I could make it happen.

DAVID: The way you did it just blew me away.

MIRIAM: How did you even know where to begin?

TARA: Well, I didn't! I didn't even really know what a non-profit was. But I knew that first we had to check to see if the public would support us.

DAVID: So Tara had the brilliant idea to send out a letter to all the addresses of my viewers. You see, over the years, every time I got a letter – this is before emails – I answered them.

TARA: You know he was on *The Nature of Things*, *Quirks and Quarks*, *It's a Matter of Survival*, he was writing for the *Globe and Mail*, he had the top genetics textbook ... He had lots of different audiences.

DAVID: And I got cards made with my picture on them. And I handwrote a letter to thank people – because I figured if people take the time to write a letter, put it in an envelope, put a stamp on it, they deserve to get a response. I had about 25,000 of these letters, and I never would have thought to do this –

TARA: We sent out a letter to all of those people, saying, 'If we started a new organization, would you support it?'

It was just before Christmas. We carried the last big box of envelopes to the downtown post office, it was midnight, and we got them sent ... but then I didn't think about what would happen afterwards.

And then before I know it, the mailman comes huffing up the stairs, and he was like Santa Claus, he was carrying this big bag of stuff, and we emptied it out, and it was all these letters. And so, the first envelope we slice open, two twenty-dollar bills drop out. The next, a hundred-dollar bill! I just thought, what does this mean? We just wanted a yes or no.

People poured out their souls in the letters they wrote to us! We even got a cheque for $1000 from this little old lady.

She said, 'I had to go to the food bank last month, but I think I can manage this, so here you go!'

I had to send it back. I felt bad about it.

I said, 'Listen, I'm so sorry, but you need it more than we do.'

DAVID: And she even wrote back and insisted! You had to write back again!

TARA: Well, she said, 'Okay, you know what, I'll send you a hundred dollars a month, and if I have to go to the food bank, I won't send it to you.' So I said okay, and she sent $1,200 that year!

STURLA: She won that battle.

DAVID: The response was just amazing! People were so full of hope!

TARA: And I didn't know what I was doing – all these people were sending us money, and I had no way to track it. I started having these nightmares of being out on a limb, sawing off the branch ... and I was on the wrong side of the cut! I was actually terrified.

MIRIAM: How did you get your feet under you?

TARA: The first thing I had to figure out was how to track every penny, and we didn't have the software to do it, and I remember one night my friend Vicky was with me, she was a volunteer at that point, and she finally said, 'Oh Tara, why don't you just give up?' And I said, 'Well, if we could just get this software problem sorted out, then we could save the world!' And it was so funny, so ludicrous that, we just fell off our chairs howling with laughter on the floor at 3:30 in the morning, thinking how pathetic it all was!

Anyway, I spent years bashing my head against wall after stupid wall, trying to figure out what job needed to be done next ... and figure out how to do it.

But that's what it took.

DAVID: How could my dad not fall in love with her?

INTERLUDE
A WALK IN THE SPRING

A poem by thirteen-year-old David Suzuki, written in 1949

Let us take a walk through the wood,
While we are in this imaginative mood;
Let us observe Nature's guiding hand,
Throughout this scenic, colourful land.

Along a rocky ledge there dwells
A fairy with her sweet blue-bells;
Singing and dancing through the day,
Enchanting all things in her delicate way.

A brilliant blue-jay scolds a rabbit,
Lecturing him on his playful habit.
A lovely butterfly flits through the air,
As though in this world it hasn't a care.

The many birds give their mating calls,
Lovelier than the Harp in Tara's Halls:
A wary doe and her speckled fawn,
Creep silently along on their moss-covered lawn.

Water cress line the banks of a stream
That is the answer to a fisherman's dream;
Teeming with trout and large black bass
That scoot for cover as we noisily pass.

The V-line of the geese reappear,
Showing that spring is actually here;

The swampy marshes are full of duck,
In the water and on the muck.

The air is filled with a buzzing sound,
From above and from the ground:
The air is heavy with the scent of flowers,
Of new buds, and evergreen bowers.

This precedes Nature's endless show,
Of all things, both friend and foe,
Living in her vast domain,
And under her wise rule and reign.

Thus within her kingdom lies,
Filling scenes for hungry eyes;
Also treasures of this natural world,
Which, if watched carefully, will be unfurled.

SCENE 3
SCIENCE + ART / LEFT + RIGHT

STURLA: Growing up in Norway, me and my friends would spend all day outdoors – we'd just point in a direction and go! We'd go into the mountains for hours, we'd take a little boat out into the sea and discover islands, we'd pretend to be explorers!

DAVID: That's the thing, you know, when you're a kid you're not aware that you love nature. You're just outside, you're playing, and as a kid that's where I was happiest too. I collected insects, especially beetles, and I LOVED fishing.

STURLA: Me too! And free diving!

DAVID: You free dive!? Wow, I would love to do that one day!

STURLA: Come to Bergen! I will take you to the best spot to pick scallops all day! But we'd better hurry – you're eighty-five now!

MIRIAM: When I was a kid I loved to swim, so I had dreams of becoming a marine biologist.

TARA: So why didn't you?

MIRIAM: Well, I thought I would spend all day swimming with the whales and the dolphins, but when I learned about all the sciency stuff I'd have to do – all those big words I'd have to memorize – I think I got intimidated. And then I fell in love with theatre and that was it!

DAVID: You know, that's how it is when you find something you love! I was studying to become a medical doctor. But in my third year of college, I had to take a genetics course, and it changed my life. I was already into biology, which was more of a descriptive science – you know, we'd collect and classify animals, memorize internal organs, or count numbers of birds in trees.

STURLA: We would do that too as kids on our expeditions – like Darwin!

DAVID: Exactly. It's thrilling to observe nature like that, but then for me genetics was a whole new world in which I could do more than just observe nature. Through experiments, I could dive deeper and actually see how nature worked! It was a mathematically precise and elegant way to probe nature.

MIRIAM: What do you mean by 'probe nature'?

DAVID: Well, you could breed the flies, and from their offspring you could uncover the laws of nature governing how genes are distributed! So genetics was a way of really probing into the structure of inheritance – like a detective story. And fruit flies gave me the 'clues.' Those fruit flies were like flying bags of chromosomes that we could analyze through genetic crosses – and we applied a lot of that learning to understanding heredity in other species – including humans.

TARA: But it was more than that – your heart was in it. You loved the flies, you thought the flies were beautiful. He was like a poet and didn't know it. He was always talking about their gorgeous eyes.

DAVID: So I often worked in the lab till two or three in the morning counting flies. Many times the janitors would come to clean the floors and wonder what this funny guy was doing ... I'd call them over to take a look through the microscope, and they were always blown away by the beautiful bright scarlet colour of the eyes –

TARA: See what I mean?

DAVID: And then when they looked further they could see the arrangement within the eyes, see their facets – each one is a kind of light-receiving organ, all arranged in symmetrical rows, and where they joined there was a little hair. And that's the spirit that made me fall in love with science – that sense of wonder!

STURLA: That's not how I think about science.

MIRIAM: It's all technical terms –

STURLA: Or things that explode!

DAVID: That's the trouble. If I write a scientific paper that is full of that wonder and say that 'I am enthralled with the beauty of the scarlet colour, and amazed at the geometric organization of a fly's eye,' it would never get published! Science is presented as an objective exercise, so we scrub the emotion out of it, yet it's that emotion and the JOY of discovering nature that attract us to science in the first place. I think it's really tragic that we do it that way.

STURLA: Yeah, I never thought of it that way. So it's actually a problem of language. 'Cause the curiosity to understand the world and how it works is the same between artists and scientists.

MIRIAM: Yeah, for me, art is a way of making visible what is invisible. In the same way you do, David, when you look into a microscope and reveal the invisible world inside – it's poetic.

DAVID: Yes, it is poetic, but in science you're supposed to hide that.

MIRIAM: Yeah, I hadn't thought of you as an artist before but you are one!

DAVID: No no, I'm the scientist – she's the artist. She loves poetry –

MIRIAM: Yeah, you taught literature!

TARA: Yes, and it's funny, I actually fell in love with literature because its great subject is nature, and the Romantic poets are a terrific example. 'O wild West Wind, thou breath of Autumn's being ... hear, O hear!' It's all about nature.

DAVID: Longfellow. 'I wandered lonely as a cloud – '

TARA: Darling, that wasn't Longfellow, that was Wordsworth.

DAVID: Wordsworth, right. My dad used to cite that all the time: 'I wandered lonely as a cloud that floats on high o'er vales and hills ... When ... something or other ... '

TARA: 'All at once, I saw a crowd ... '

DAVID/TARA: 'A host of golden daffodils!'

TARA: And I think it was Wordsworth who predicted that in the twentieth century the great epic subject of literature would be science, and the scientist would walk hand-in-hand with the poet. But it didn't happen! And I was desperate to know why. That's why I wrote my PhD thesis, because I wanted to know why.

STURLA: Why art didn't embrace science?

TARA: Yes, exactly.

DAVID: Get your pen out because this is brilliant, put this in the play – we need to get this published!

TARA: So my thesis went something like this: our brains are divided into two hemispheres, a left and a right brain. But they're not duplicates, they each perform different roles.

Both hemispheres are important and they communicate with each other through a huge band of tissue called the corpus callosum.

So I'm aware I'm oversimplifying – and the brain is plastic, and highly adaptable – but the left brain is linear, it's directional, it deals with measurement of all sorts, numbers, it breaks things down into parts. It's our analytical and logical side. It thinks of people as individuals rather than communities, for example.

STURLA: Got it.

TARA: Whereas the other brain, the right brain, it's the world of imagination. It sees time as cyclical. It's pre-linguistic, it's the world of the unconscious, of dreams, music, dance, and feeling. It sees everything – nature, humans – as interconnected.

MIRIAM: So there's reason and imagination.

Reason, the left side, breaks things into parts, science. Imagination, the right side, sees the whole, art.

TARA: Exactly. So the left side of the brain can see two eyes, a nose, and a mouth. But it takes the right brain to recognize a face. The point is, we need to use both sides of our brain to thrive – we need a balance.

DAVID: For 95 percent of human history, we had to have a balance of left and right brain thinking because we were hunter-gatherers. People understood that we were utterly dependent on nature.

We were embedded in a web of relationships with other life forms that we depended on for survival.

MIRIAM: So when did that change?

TARA: Well, that lasted up and into the Renaissance. Do you remember the Renaissance men?

STURLA: Like da Vinci?

TARA: That's right. He was a scientist and an artist. That society valued a balanced mind that develops all its faculties – both left and right brain.

But over the last two centuries, science, the left brain, became more valued as the superior mode of thought. And then (fast forward many years later), during the Industrial Revolution, technology gradually became more and more powerful.

DAVID: With technology, we had telescopes that let us see to the edge of the universe, microscopes that revealed a world of life in a drop of water, machines that could work twenty-four hours a day that could lift and move enormous weights. We were no longer limited by our biology because we could defy gravity, we could fly as fast as we wanted, dive as deep in the ocean as we could imagine. Nothing could limit us except our imagination.

TARA: We used to say 'Faith can move mountains,' but suddenly technology *really* could and did! Nietzsche declared 'God is Dead', and it became hard for people to disagree that scientific left-brained thinking was superior. So poetry and the arts became pushed to the side, and as we marched into left-brain thinking, we separated ourselves from nature.

DAVID: This is the heart of the issue, the root cause of our destructiveness toward the planet. We fail to see ourselves as part of an intricate web of relationships. We've even killed God.

TARA: That's exactly what Nietzsche was saying, here's the full quote: 'God is dead ... And we have killed him.

'How shall we comfort ourselves, the murderers of all murderers? What was holiest and mightiest of all that the world has yet owned has bled to death under our knives: who will wipe this blood off us? What water is there for us to clean ourselves?'

STURLA: Oh wow, that is powerful. The image of God, or the earth bleeding to death with a knife in my hand ... it gives me chills.

TARA: Yeah, it makes you *feel*. It touches your heart.

DAVID: The environmental crisis is not just a scientific, logical issue we're confronting. We've got to touch people's hearts.

TARA: We've always wanted to learn how to communicate science in a way that changes public behaviour, so it is communication that is the really hard nut to crack.

DAVID: Thirty years ago we stopped some logging, stopped some oil wells, stopped some dams. And today we are fighting the very same battles we thought we had won. The damn dams are back! So we didn't solve the issue. We hadn't found the source of the problems. We didn't shift the perspective.

MIRIAM: So how do you shift perspective?

TARA: Good question. Well, we've gone so far down the road of finance and numbers and straight-line progress and practical picking things apart that we have forgotten about synthesis – putting things together. David and I had the privilege of learning from our Indigenous brothers and sisters, like Paiakan in the Amazon, Miles Richardson and Guujaaw in Haida Gwaii, or Pauline Waterfall in Bella Bella; they teach us the earth gives us all our nourishment, creates our bodies – the earth *is* our mother, everything is related, everything is our kin.

DAVID: To me, as a scientist, this all sounded like New Agey stuff until I was shooting an episode of *The Nature of Things* on Haida Gwaii in the late seventies. There was a group of the Haida people protesting the logging in Windy Bay. And I remember interviewing Guujaaw (who ended up becoming the president of the Haida Nation and a dear friend), I said, 'Well, you're an artist. What difference does it make to you if they cut down all the trees?'

And he replied, 'If the trees are gone, we'll still be here ... but then we'll be just like everybody else.'

And something clicked. He was telling me we don't end

at our skin or our fingertips. The rivers, the birds, the trees, the fish, and all of that are connected and make the Haida who they are. And if you start thinking about it, well it is true, you know, it's the air, the water, the fish, the plants, that make our bodies, like we literally are created out of the earth, the earth is our mother.

TARA: That's why Guujaaw had to fight for the forest – he understood that he was intimately connected to the earth – to everything – and that meant he was responsible for protecting it.

STURLA: Imagine how we would live our lives if we all grew up hearing the story that we are interconnected to all living things.

MIRIAM: It's easy enough to 'understand' but to actually *feel it* is a whole other thing.

TARA: To see ourselves as a part of nature means we have to embrace what Indigenous people have known for thousands of years, which is that we have a spiritual connection, a sacred relationship to nature, and that requires both left and right brains to be in balance. I feel I really understood it in my time in the Stein Valley ...

MIRIAM: What happened there?

TARA: I'm not sure I could describe it with words. It was through ceremony. And ceremony – like poetry – is a portal to the right brain. It was ritual. I was listening, involuntarily, to a drum beating outside this teepee we were sleeping in, all night long. At first it was irritating and then something clicked ...

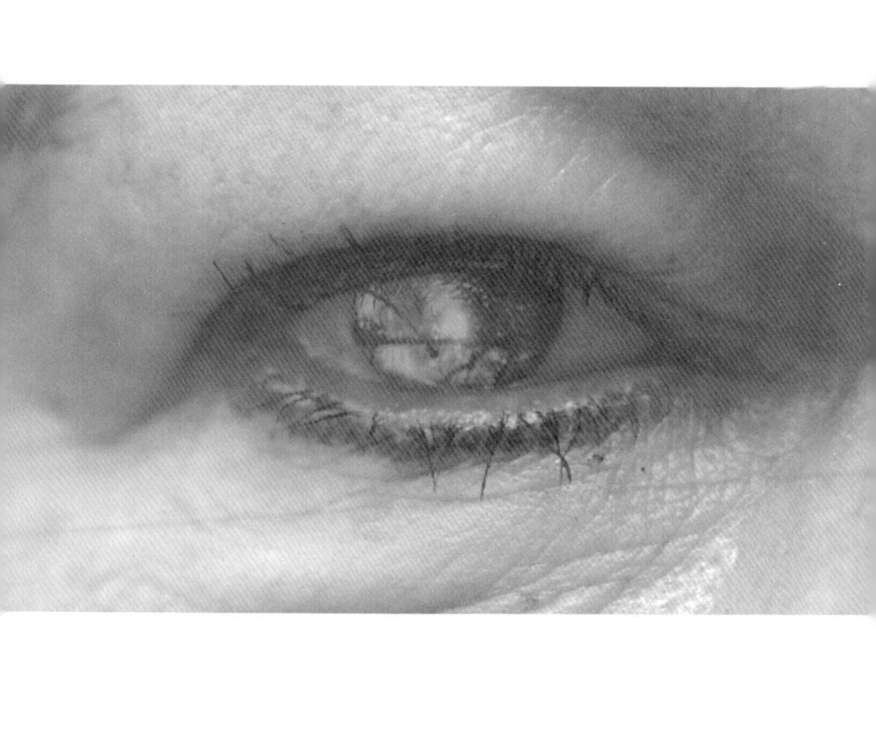

INTERLUDE
STEIN VALLEY

I – I wasn't really prepared to be uncomfortable.
I thought it was going to be ...
We came as strangers ...
I had my little girls – they were six and three.
And they showed us to where we were to sleep, which is in a teepee.
I had a nice sleeping mat for the children, to keep them comfy, 'cause they need their sleep.

And then, just as they were settling down to sleep, I realized that someone was coming up beside the tent. And I realized that they were carrying this heavy thing and they put it down and it was a big drum, and then lo and behold, they started drumming.

And I thought, don't they know we're in here?
I couldn't understand –
And I was so worried about how to handle this
And I just felt so confused
And it went on and on and on ...

I could hear the drums. I could hear the singing.
I could hear the running of the water of the Stein.
And I could just ... it was just so many things in my head. It was dark, but I knew there were so many stars in the sky. I had never seen it like that before. And then there was just this sound of the singing and the repetition of it, you know, repetition ... repetition. And then there was the drum, right, and, well, after a while it just gets into your heartbeat.

It's like I'd gone through a portal to a different world, really. Yeah, it changed my life.

SCENE 4
THE AMAZON

DAVID/TARA: Itadakimasu.

MIRIAM: What does that mean?

DAVID: It means, 'I am going to receive the lives of animals and plants for my own life.' Saying this phrase before eating is a way to express your understanding of how much was sacrificed to make the meal possible as well as to express appreciation for Mother Nature.

MIRIAM: That's beautiful.

TARA: And now we add, 'Thank you, fish, for giving us your life.'

STURLA: Mm, this looks so good. So, David, you were starting to tell us about Paiakan.

DAVID: What was I saying about him?

STURLA: You were just telling us how you met.

DAVID: Oh, right. Well, meeting him changed my life. I met him when I was filming *The Nature of Things* in the Amazon in the late eighties.

TARA: This is when people realized that the Amazon was the lungs of the planet, and the fact that it was burning made a lot of people upset.

DAVID: Yeah, so we're down there filming this special, and when Paiakan and I meet, it's this magical moment. It was like we had met before – we had an instant connection.

TARA: Yeah, he really took to you. It was like he recognized you.

DAVID: I learned later that when he was a boy, in his village way in the forest, the village used to be visited by a Japanese-Brazilian doctor who would come to treat the people. *And* it turns out his name was Davi, which is Portuguese for David.

TARA: And that doctor, Davi, became a kind of mentor to Paiakan – he told him, 'You learn to protect your people, the white people are coming'. And when the village, called Gorotiri, finally got a road into it, sure enough, Paiakan saw alcohol, white bread, and candy coming into the community and he said, 'This is no good, we have to leave.' So he led a group of people further into the forest, where there was no road, and founded a village called Aucre.

DAVID: And when he saw that I was Japanese, then heard my name was David, it was a kind of reunion.

TARA: They really had a special connection.

DAVID: In a very short time, we became quite close. And he confided in me that he was shocked to find out from an Oxford anthropologist –

DAVID/TARA: Darrell Posey –

DAVID: That the Brazilian government was planning to build a dam that would flood Kayapo territory and they weren't even told about it!

TARA: That doesn't just happen in developing countries. That's exactly what happened in BC when the Bennett Dam was built: the Indigenous people weren't told about it, and the first thing they noticed was the water was rising in the river and beginning to flood their graveyards. And the caskets began to float down the river!

DAVID: The same thing in Quebec: by chance, Phil Awashish was in Montreal, picked up a paper, and found out that Hydro Quebec was going to flood Cree territory. So when Paiakan said to me, 'I need help', I knew exactly what he meant. That's when I called Tara.

TARA: Yeah, David phoned me from the middle of the Amazon, and he said, 'The forest is burning, it's so smoky our plane can't take off. Tara, you've got to do something.' And he had this catch in his voice – you know, I could hear that catch 5,000 miles away, and listening to it changed my life – so I said, 'Sure, okay, um uhhhhh, what?'

DAVID: So, I said, 'Paiakan has this vision of bringing all the affected tribes to where the dam is going to be built. He wants to gather everyone to tell the government, 'No dam!'

STURLA: So what did you do?

DAVID: We brought Paiakan to Canada to raise money for him, so he could organize the first meeting of all the Indigenous people in the lower Amazon.

STURLA: That sounds like a whole other adventure!

TARA: Oh, wow, there are so many stories to tell. Well, we hosted a big event at St. Paul's church on Bloor East in Toronto to raise money and public awareness for what was going on in the Amazon. And people were lined up around the block.

DAVID: We packed the place!

TARA: Margaret Atwood read a poem.

DAVID: The Nylons, who were a hot singing group at that time, performed.

TARA: Gordon Lightfoot joined us!

DAVID: And then Paiakan got up and talked through an interpreter, it was spectacular.

TARA: It was electric. And right before that, we had a cocktail party at the Elmwood Club. A thousand bucks a glass!

DAVID: This was Tara's brilliant idea. She said there are a lot of companies down in the Amazon making a lot of money –

TARA: Canadian companies –

DAVID: – so let's call them and say they should come, because Paiakan is coming up to speak. Now, a lot of them didn't come but still paid the thousand bucks.

TARA: And CBC filmed it. There was a red carpet, and they all got to meet Paiakan, and Margaret Atwood and Gordon Lightfoot, it was really successful. And then we all got into cars and drove to St. Paul's and saw that lineup around the block – it was sold out!

Our jaws dropped, we couldn't believe it. And inside, everybody felt they were part of something absolutely historic! They knew we were doing something positive for the Amazon. Then we did the same thing in Ottawa – Elizabeth May helped us do that.

DAVID: We raised a total of $75,000.

MIRIAM: That is a lot of money.

TARA: And don't forget this is 1988.

STURLA: Yeah, a car cost a nickel back then, right?

TARA: Haha, yeah. Anyway, so we got the front page of every paper in Toronto, Ottawa, and Montreal because of the regalia that Paiakan wore – it was so colourful and dramatic, you know, the feather headdress and black paint across his face. His face was everywhere on the newsstands!

DAVID: What I loved was, we had this press conference, and this one cheeky guy said, 'Why do you wear all this paint and feathers?' and Paiakan didn't miss a beat, he said, 'Why do you wear a necktie?'

TARA: Suddenly there was so much attention on that dam that even the Canadian government couldn't ignore it. And Michael Wilson –

DAVID: Who was Minister of Finance at the time –

TARA: He said he had more phone calls and letters on that issue than anything else during his tenure in Ottawa. And he decided Canada would vote against the World Bank's loan that was being used to build the dam. That was three and a half percent of the vote! It doesn't sound like much, but it was really significant because it influenced other countries as well. But there was still a lot of work that needed to be done.

MIRIAM: Like what?

TARA: Well, we needed to find a way to get that money down to Paiakan – he had returned to Brazil, so it wouldn't be easy to get it there.

DAVID: And we couldn't just wire it down.

TARA: There was no internet banking back then.

MIRIAM: Don't tell me you just carried $75,000 cash down in your purse ...

TARA: Well, no – there's another story there – but we actually did have to cloak-and-dagger all that cash down there. So not only was that complicated and a bit dangerous, but, well, on top of that, Paiakan wanted to hold a meeting of the Indigenous peoples of the Lower Amazon – in the rainy season! It was an incredible feat of logistics! But Paiakan got it done.

DAVID: He had to organize buses and boats to go down those rivers –

TARA: He organized it all, and set it all up, so these tribes could all meet at a town called Altamira, which is near the mouth of the Xingu River where the dam was to be built. And as all this was happening, I said, 'David, we've worked so hard to raise that money, we've got to go and make sure that it works out and the World Bank turns it down.' I wasn't going to miss that for anything!

DAVID: So I say, 'Great!' and then next thing we know Elizabeth May wants to come with us, and Gordon Lightfoot wants to come with us, and we ended up – Tara ended up – with forty people!

TARA: A gang of forty people and I had to be the travel agent –

DAVID: She did it all, it was incredible.

TARA: Well, it was quite the adventure, organizing forty people on tiny planes flying in remote towns in the Amazon –

DAVID: And she did it all in Portuguese!

MIRIAM: You speak Portuguese?

DAVID: She learned it for the trip!

TARA: Well, yeah, someone has to – how else would anything get done?

So we finally get to Altamira, and Paiakan has assembled all of these Indigenous tribes – dozens of different tribes, all in their own regalia – and there was the mayor of Manaus, the press is there, a rep from the World Bank, the head of the energy company Eletronorte that was building the dam (he was a brave guy), and representatives from the Brazilian government –

DAVID: (*Privately to Sturla*) You know Sting?

STURLA: Yeah.

DAVID: He was there.

STURLA: You know Sting?

TARA: 'Eshtingy' they call him. And everyone is packed into this big hall and it's so exciting – it is also scary as hell because the Brazilian soldiers are surrounding the room, and they have their great big assault rifles, and every time something dramatic happens, they tense up, you know, and then we tense up, and we didn't know how we were going to get out of there alive!

DAVID: Especially when that older Kayapo woman ran up to the front, berated the Kayapo warriors for cowardice, then spun around, lifted her machete, and with the flat side of it, slapped the Eletronorte representative across the face!

MIRIAM: What are you talking about?

DAVID: Twice!

TARA: Paiakan was the MC. He grabbed the mic and calmly told the room, 'We Kayapo are a theatrical people.'

MIRIAM: Guns and machetes and you weren't terrified?

DAVID: I was shitting bricks!

TARA: You know, fear is always there, but if you let that stop you, you won't get anything done. You know, I think it was just ... it was everything I ever wanted to do, I think it was the most fun I've ever had.

MIRIAM: So what happened in the end?

TARA: After all of that, the World Bank withdrew their promised loan of $500 million. And that killed the dam dead!

MIRIAM: Whoa! So you stopped it! You won!

DAVID: Well … for thirty years, it was stopped in its tracks, but now the project has been revived. In 2016, the dam was begun.

STURLA: What? How could that happen?

DAVID: You know it's again one of those things where we fought and won, but it was only a skirmish. We didn't win the war. I mean, we stopped the proposal to drill for oil in the Arctic National Wildlife Refuge, we stopped the dam at Site C on the Peace River. But now most of those very same projects have not only been revived but are going full tilt a generation later. So we may have won the battle in the Amazon, but not the war. We didn't change people's minds.

MIRIAM: Right. If the underlying values and beliefs that drive us don't change, we won't *really* change –

DAVID: Exactly.

TARA: The point is that if you stop oil from being pulled out of the ground, it's still there and can always be extracted later. Every generation needs to understand it's responsible for taking care of the planet for the future.

DAVID: The real war isn't to stop the dam or leave the oil in the ground, it's to change people's hearts and minds.

STURLA: To change the story …

TARA: Yeah! To see everything as interconnected.

DAVID: And that interconnectedness involves a reciprocal agreement with nature.

MIRIAM: It's like you both said earlier: the earth is our mother – she gives us shelter, she feeds us, protects us from harm. That's love – and no loving relationship is sustainable when it goes only one way.

DAVID: Exactly. This is something that must be understood: that we live within a web of relationships with nature – with the air, the water, the soil, the sunlight. Since time immemorial, Indigenous people around the world, through their ceremonies and rituals, thank their creator for nature's generosity and promise to honour a reciprocal obligation – to act properly, to keep it all going. And that reciprocity is what's missing in modern society.

TARA: It's gratitude. If we acknowledged the love we received more often, then we would feel responsible to give love and appreciation in return.

MIRIAM: Itadakimasu.

DAVID: Right. Itadakimasu.

INTERLUDE
ON IMAGINATION

Excerpt from 'On Imagination' by Phillis Wheatley

Imagination! who can sing thy force?
Or who describe the swiftness of thy course?
Soaring through air to find the bright abode,
Th' empyreal palace of the thund'ring God,
We on thy pinions can surpass the wind,
And leave the rolling universe behind:
From star to star the mental optics rove,
Measure the skies, and range the realms above.
There in one view we grasp the mighty whole,
Or with new worlds amaze th' unbounded soul.

SCENE 5
FEMINISM + THE ECONOMY

TARA: Let me tell you a story. Say you live in a little house, and the wife – or the husband for that matter – stays at home to care for her children. She makes pies and hands them across the fence to the next-door neighbour. She volunteers at the local daycare, at the old folks home, and she takes care of her own aging parents upstairs. She supplies all kinds of baked goods for every fundraiser at the children's school. But she's not contributing to the economy because it's all just over the fence and doesn't involve money.

Well, now she gets a divorce. So the lawyer gets paid a lot of money. Both she and her spouse have to go to work so the government gets more taxes. She can't bake pies anymore for the neighbours, so the neighbour has to go out to the store and buy his own pies. He can't drive so he takes a taxi. And the GDP goes up.

The wife has to sell the house. Her parents upstairs have to pay to go into a home.

MIRIAM: The GDP goes up.

TARA: Exactly. Now, she falls ill, and so that's great – for the hospital and the doctors. There's an ambulance. She dies. There's an undertaker.

MIRIAM: There's a funeral home. A florist.

TARA: And The GDP goes up and up and up. And meanwhile, the school can't get those donated goodies, so they have to purchase them in a shop. The basketball team she used to coach can't afford to pay for a coach and falls apart. All of the places where she used to volunteer have to start paying for those services.

MIRIAM: So, as society breaks down, the economy grows.

TARA: It's like it feeds on society. The GDP goes up, and all of the work done to care for parents, neighbours, the children and community – it doesn't count.

STURLA: Oof.

DAVID: Do you know Adam Smith, the father of modern economics?

MIRIAM: Yeah, I studied him in school – *The Wealth of Nations*, right?

DAVID: Right, well, he was talking about all of the factors that contribute to the economy and he absolutely ignored women.

MIRIAM: Classic.

DAVID: Meanwhile, he's living with his mother, his mother is doing all of the cooking for him, and washing all of his clothing, and all of that stuff that allowed him to sit in his room and write his magnum opus, and yet his mom's labour was completely irrelevant!

TARA: It wasn't accounted for. It was an externality – like nature, it was invisible to him.

DAVID: He was known for writing about the invisible hand. The real invisible hand is the hand of women.

MIRIAM: Hear hear!

TARA: Sure, but he never talked about the invisible hand of women – that was David! Adam Smith got it all wrong! He said the basis of economics is self-interest. I think that women's traditional role in society has been this fight against the entropy – the chaos – of self-interested capitalism that is tearing everything apart. But the antidote to that chaos, the opposite power, is love. It takes love to create children, to keep them safe, to make meals, to make a household and keep it from exploding from all of the external forces, and to keep the community together and alive. Love is the opposite of self-interest, and it was invisible to Adam Smith.

DAVID: It's work – it's hard work.

TARA: It's sheer, goddamn work, it tears your bones and sinews out of you. And the annoying thing about it is that the greatest success you can have as a housekeeper at the end of the day is that the house looks the same as in the morning. There's no salary, there's nothing to show for it except a lack of mess. So it is the most unrewarding work. Right, Sturla?

STURLA: I know what you mean.

TARA: And I often think that environmentalism is global housekeeping.

If you spill oil in the hall, every housekeeper (man or woman) knows that if you don't clean it up right away it'll get trodden all through the house, into the carpets and into the sheets ... it takes a lot of work to care for something, and if you stop, it falls apart. And the greatest thing you can hope for at the end of the day is that the valley still exists, the forest is still standing, the water is still in the lake, *and you can still drink it.*

DAVID: It takes a lot of work to keep fighting hard like that.

TARA: And that is love. When you love something, you'll do anything to protect it.

(*Motioning to Sturla, who is cleaning the table after dinner*) Right, Sturla?

STURLA: I know what you mean.

SCENE 6
THE KISS

DAVID: When I was in Grade 13 – this was in 1953 – I had a couple of guys that were friends – we were all nerdy guys like me – and they said, 'Hey, why don't you run for class president,' and I said, 'Hell, no. No, I wouldn't win.'

So I'm telling my dad that night, 'You know that Vic, he's so stupid – he wanted me to run for school president. Isn't that crazy?'

Instead of laughing, my dad says, 'Well, why don't you, why wouldn't you run?' I said, 'I don't have a chance!'

And then he got really mad and said, 'How the hell are you going to know? There is no shame in trying and losing. There will always be people who are bigger, stronger, better than you, but how will you know what you're capable of if you don't try?'

STURLA: That's beautiful, and so true.

TARA: And he won!

DAVID: I never expected to. I now know that back then the high-school world was divided into two groups: the Innies and the Outies. The Innies are basically the beautiful people, you know; they set the social scene.

STURLA: *(Motioning to Tara and himself)* So, this side of the table.

DAVID: They play football or basketball or are cheerleaders. The way they dress sets the style for the school. And the Outies, are everybody else –

TARA: Which is most people.

DAVID: And, back then, if you were a 'brain' – a brain was someone who got good grades – that was one of the worst things you could be. You know, it was like having leprosy or a venereal disease, it was a pejorative to call someone a brain. And I was a brain!

So I ran on the basis of 'I'm not one of the Innies, I don't play basketball, but I'm committed to these ideas of what the school should be.'

I ran as an Outie –

TARA: And he got more votes than all of the others put together.

STURLA: There are always more Outies than Innies ...

MIRIAM: So true!

DAVID: Yeah, and that whole concept, that there is no shame in trying, that was really an important lesson to me.

STURLA: Oh man, I learned that lesson too – the night I finally professed my love for Miriam! It was the bravest moment of my life.

MIRIAM: Oh yeah! We had just gotten back to Paris.

STURLA: The city of love!

MIRIAM: Yeah, we were at school there, and after being apart for the summer, I thought we were just going to stay friends ...

STURLA: But the moment I saw Miriam again after the summer, it hit me so hard ... you know? The love. So I went home that night and I couldn't sleep. I was lying in bed, turning, sweating, tortured by love, a love that unless I was able to tell her, would torture me for the rest of my life as regret.

DAVID: Well, of course! You've got to tell her, there's no harm in trying!

TARA: And think about what you would have missed out on if you didn't ask her!

STURLA: Right! So eventually I managed for the two of us to be alone in a street cafe in Paris, and I declared my love for her then and there – she had no idea what was coming!

MIRIAM: I was shocked! I had asked him to help me move apartments – so we're sitting at this cafe surrounded by all my suitcases – and I've never seen him like this before ...

STURLA: I was so nervous! I was just saying a lot of crap, I was stuttering, and then I had a moment of clarity, and I said, 'When I tell stories to my friends and family and grand-children in the future, I want you to be the protagonist in all my stories, and I want to be the sidekick in yours.'

TARA: Oh, that's so poetic. Miriam – what did you say?

MIRIAM: I'm less romantic – I said I needed time! Like you, Tara, I wasn't in a rush.

STURLA: Well, I went home and slept like a baby. I had done everything I could, and I had no more to regret. At least I gave it the chance to either fail or succeed. It was kind of out of my hands.

TARA: How romantic – it sounds like a Norwegian film!

MIRIAM: Yeah, except in a Norwegian film nobody would say anything. They would just look at each other across a fjord longingly –

STURLA: Well, we would understand love in the silences ... and ... maybe we are a little emotionally repressed ...

TARA: It's not always easy to be articulate when it comes to love.

STURLA: Yeah, the feelings are so strong that my brain doesn't work, like it can't process the feelings into words.

TARA: And sometimes words just won't do it. Like with David, he just asked me to dance and then ran away!

MIRIAM: What? That's how you asked her out?

DAVID: Yes and no. I was giving a speech about how this beautiful science – genetics – has been misused to justify genocide.

MIRIAM: Wow! Nice line – that sounds very romantic.

STURLA: You led with genocide and then you asked her to dance?

DAVID: Oh, actually, this is a great story. You've got to tell them. It was 19–

TARA: –71. And I'm a Master's student at Carleton University. Well, I should take a step back.

So you should know my dad trained me to think you don't go looking for boys, you aren't supposed to be interested. So I got all the way up to the last year of my Master's only really being with the people who wouldn't give up. So I didn't really choose my own boyfriends, they were just the most persistent ones.

Then one day I had this realization. I thought, I really work hard on every essay that I write, I put my guts into it, you know. You'd think I'd lift a finger to choose who my life partner might be. And I'm finishing my Master's this year. There are thousands and thousands of interesting people around my age here at the university, but once I get out into the world, it'll be like my temp jobs in the summertime: there'll be five people, two of whom are men, neither of whom I would want to have a cup of coffee with.

So, I really ought to make a little bit of effort. Okay, Cullis, this is a project, what are you going to do? I thought I'd better go somewhere I might meet some interesting people. Where's that? Hmm, maybe the grad-student pub. Okay, off you go. Met this nice fellow, went back to my roommates and said, 'Hey, I met a guy tonight, I'm thinking I might see him again!' 'Oh, um, what's his name?' 'Tom Smith.' 'Uhh, did Tom Smith tell you about his wife and three children?'

So I thought, holy smokes, this is going to be harder than I thought. I've got to put more effort into this.

So I said to myself, look, if you come across somebody who is interesting, don't do what you've always done, which is think, 'Gee, I hope our paths cross again one day,' but get in front of that person and meet that person, don't just let it slide. Put some effort in!

Just two weeks after I made that decision –

DAVID: It was December 10th –

TARA: 1971. I was teaching a class in the arts tower at Carleton, and I wanted to go to hear David Suzuki speak – he was speaking at lunchtime, and I was homesick for Vancouver and he was a Vancouver boy, so I thought it'd be nice –

DAVID: I never went to these lectures to pick anyone up. You know, I had achieved a certain amount of notoriety because I had this science program –

TARA: *Suzuki on Science* –

DAVID: And for me it was just another lecture.

TARA: My students were kind of needy that day, and they hung around after class and asked so many questions, it got late, and I thought, ah, what's the point, but it's only two floors up. So I put in a little effort, I went up the stairs and squeezed into the room – the room was so full –

DAVID: I thought you were in the front row –

TARA: I couldn't have been because I came in late.

DAVID: But I mean, there she was, I noticed her right away, and it was like there was a beam of light shining on you. I mean, you were gorgeous. Honestly, she looked just like Rita Hayworth. You know Rita Hayworth?

MIRIAM/STURLA: No./Yes.

TARA: Now, he's very personal when he gives a lecture. He talks about his own experiences – you do talk about yourself a lot – and so I learned a lot about him. It gave me a lot of insight into who you were, what you were like, and I thought, well blow me down, that's the first person I've ever seen who I would consider marrying.

STURLA: Nicely done!

DAVID: Works every time!

TARA: And that was after an hour and a half of lecture. That's a long lecture ...

DAVID: And toward the end of my lecture, I realized that suddenly people stopped paying any close attention to what I was saying. So I looked over at what they were watching, and there was this immense white owl, an arctic owl –

TARA: Snowy owl –

DAVID: Yeah. And this is the highest building on campus, right? And it had flown right up on that level, and hovered, looking in the big window.

TARA: It was striking, I'd never seen anything like that. This great, HUGE white owl. I didn't really know how big they were!

DAVID: So I finish up the lecture –

TARA: And then he said, 'Does everybody know about the party tonight?' and I could have sworn he said that for my benefit.

MIRIAM: (*Turning to David*) Did you?

DAVID: I did!

TARA: So I went home and washed my hair, 'cause that's what you do. I asked my first roommate, 'Will you come with me to go back out to Carleton for something that's going on tonight?' And she said, 'No, I'm going to the Gatineau for the weekend.'

And so I went to my second roommate and I said, 'Well, would you come with me out to Carleton?' and she said, 'No, I'm going to Toronto for the weekend.'

And so I thought, don't tell me I have to get on the bus all by myself, and go out there like some idiot and wander about the university to find something to do with David Suzuki, but then I thought – I'd made myself a promise, and so I didn't give up.

DAVID: Thank god you didn't!

TARA: So anyway, I got out there, I found the party. Got to this house. Got up the stairs into the kitchen and I could see David was in the corner, and all around him were all of these students just peppering him with questions and pressing in on him, you know? I remember thinking, that's it. I've done everything I can do. I had reached my limit. I thought I was really going out on a limb – if this isn't it, forget it. And so I sat on the edge of the sink, and David disappeared down, he just disappeared and popped up on the other side of those people in front of me and said, 'Would you like to dance?' And then he took off!

DAVID: I didn't want to get turned down!

TARA: So I said to the girl beside me, 'Was he talking to you?' And her boyfriend said, 'He was talking to YOU!'

But I wasn't sure ... so I went into the other room where they were dancing and it turned out I was the right person, because then David and I were dancing.

DAVID: And sparks were flying between us –

TARA: Neither of us could speak. I couldn't speak and he couldn't either, so I figured the same thing was going on in his brain. And then later that night, David dropped me home, he took me up to the door and gave me a kiss good-night, and that was that.

DAVID: That was it. That kiss was it.

SCENE 7
PASSING THE BATON

MIRIAM: So we ended up getting legally married at city hall so we could start the process for Sturla's immigration, and we were planning a big wedding in May.

STURLA: All my family and friends from Norway were going to come, and we had to cancel the wedding because of COVID.

TARA: Aw, I'm so sorry!

MIRIAM: Yeah, we'll just never get that 'start' to our marriage with all our loved ones around us, you know.

DAVID: Yeah, the hardest part of COVID for us was when the grandkids went back to school and we couldn't see them for months. It was devastating. But now with the vaccine, you know, slowly we can see other people – you're here, and the best part is we've been able to spend time with our grandkids again! I'm so happy!

TARA: Do you two want children?

STURLA/MIRIAM: Yes/No.

MIRIAM: We're not sure. Well, I'm not sure. I mean, we've gotten to spend so much time with you both, and David, you have said so many times that we're not going to make it, that it's too late.

TARA: He's always saying that about everything! That should be on his tombstone: 'It's too late'.

MIRIAM: But, David, is it really too late?

(*Pause.*)

DAVID: Tara, do you remember the day I was sitting at home, and you and Sarika were cooking away, and I was holding our grandkids, the twins, in my arms, and suddenly I started to cry?

And Tara and Sarika came over and grabbed the babies and said, 'What's wrong? What's wrong?' And I couldn't answer them. I had never felt so completely desolate with despair. And I began to wail – I just – that has never happened to me in my life ... But, for the first time, I realized that the way things are going with the planet, if we keep going down this road, there'll be no future left for them.

(*Pause.*)

MIRIAM: And you're *David Suzuki* – and when I hear you say that, I freak out! Because if we don't change now, there'll be no future for any of us!

TARA: But it doesn't have to be that way! We're not at the end of this fight, there is still so much that we can do!

STURLA: How do you stay so hopeful? Don't you get depressed?

(*Tara laughs.*)

STURLA: Why are you laughing?

DAVID: Well, it's a question we are asked over and over again, you know? *How do you keep from being depressed?* We're depressed! We get depressed, we get terrified, we get all kinds of –

TARA: We just don't both get depressed at the same time. When David is down, I'm there to pick him up, and when I'm down he's there for me.

DAVID: And you know, when you have children and grandchildren, there is no choice. You just have no choice. You have to try to make things better.

TARA: People say 'Oh, there is nothing I can do.' But I think we forget how much power and influence we actually have. And the difference we can actually make.

You know, in Canada, we have education, this is a democracy, we can get to know our Member of Parliament and vote! Sure, we do get depressed when the door shuts, but others open – or we have to do everything we can to open them! We have so much unrealized power – we just have to use our imaginations.

DAVID: To me the depressing thing is the *inaction*. You don't just give up in the frustration – you've got to do *more*. Taking action *is* hope!

TARA: What was it St. Augustine said? 'Hope has two beautiful daughters. Their names are Anger and Courage: Anger for the way things are, and Courage to see that they do not remain as they are.'

DAVID: And we have a moment here to really change things. It's not just COVID; we have the conjunction of a number of things happening. We've seen market swings to the point that the price of oil was less than the cost of extracting and shipping it. Then George Floyd was murdered and Black Lives Matter exploded, and people were losing jobs and income during the lockdown while the very rich were getting even richer! People are angry and they should be!

TARA: We have to find the courage to make a seismic shift, to see everything as a whole. We can't come out of the pandemic and deal piecemeal with issues of hunger and poverty, climate change, and racial justice. The pandemic has given us the opportunity to recognize our real values and our relationship to what is really important.

MIRIAM: So how do we make that huge shift? What story could we possibly tell that could change our perspective?

DAVID: We should remind people that for most of human existence we understood that we were deeply embedded in nature and that we relied on all the things that nature provided. That's an ecocentric view of the world.

But in modern times, we started to see ourselves as the centre of the action – you know, the whole world revolves around us. An anthropocentric view. And so the story we are told and believe is that everything is just a resource or an opportunity for *us*.

All of our religious, political, economic, and legal systems are built upon that idea – that it all revolves around us.

It makes us think only about *our* rights, *our* property, *our* ownership – but what about the rights of the lake? The rights of the rivers? The rights of the trees?

We have to dig deep into ourselves, and understand and admit that rich story Indigenous communities everywhere tell: that despite everything, we are part of nature; that we are animals, wonderful animals; that our hubris means we're playing with extinction here, and that to survive we have to love nature enough to put her first.

(*Pause.*)

Well, we could go on and on all night ... Hey, have you thought about who will play me and Tara in the play?

MIRIAM: Not yet. I was thinking we would just play ourselves.

DAVID: What about the guy from *Kim's Convenience* to play me?

TARA: Oh he's great, he'd be great!

STURLA: I don't think we can afford him!

DAVID: Okay, never mind the casting. What's the ending going to be?

MIRIAM: Well, you're the scientist, shouldn't you be answering that?

DAVID: No. You're the writer, you've got to finish the story.

MIRIAM: Tara, you're the poet. What do you think?

TARA: Well, there are two sides to this story and it is a continuous contradiction: yes, it is bleak and scary and we may not make it – and yes we are going to do everything we can to make sure we get there.

DAVID: My hope is that if humans can pull back and give nature a chance, I think she will be far more forgiving than we deserve.

(*Pause.*)

TARA: You know, the truth is it never ends. There is no end, things just transform. Every generation passes the baton to the next.

Just think of Greta Thunberg taking a day off from school. Or Paiakan: he was a young man, living in the heart of the Amazon rainforest, he was in his late teens, and he was able to make a huge difference in the world –

DAVID: And now *his daughters* are carrying on the work.

MIRIAM: And *your* daughters are also carrying on the work ...

TARA: Yeah, Severn just started as ED of the foundation.

DAVID: It seems like yesterday she was just twelve years old, starting the Environmental Children's Organization, figuring out a way to get the group to the Rio.

STURLA: For the '92 Earth Summit – I saw that speech on YouTube.

MIRIAM: 1992 ... That would have been just after you started the foundation, right?

TARA: Just about ...

MIRIAM: And now thirty years later Severn is leading the foundation you started.

TARA: Yeah.

MIRIAM: And here we are spending an evening with the both of you – all from an email I sent to David five years ago. I can't believe we made this happen.

TARA: When you take action, you never know what influence will ripple out, or where the sparks will land and catch fire.

DAVID: The important thing is that we do. The doing. That is what defines us.

TARA: I know it can be overwhelming. Believe me, I know. Sometimes you just need a reminder that you're not alone.

And you're never going to say, 'Hey, we saved the world, let's go on vacation!' It's been forty years and I'm still waiting for that vacation! So you have to love it. You have to look for opportunities to keep the electricity going, or to spark it over again. I mean that's what life is all about. Or what our lives are all about.

(*Pause.*)

DAVID: I still don't know how you're going to turn any of this into a play ...

MIRIAM: I've got it. It's a love story.

TARA: Oh, a love story?

MIRIAM: Yeah, but it's not about us ...

DAVID: I don't get it.

MIRIAM: We're not the centre of the story ... Nature is.

DAVID: The guy from *Kim's Convenience* is going to play Nature?

MIRIAM: No no, we're still in it –

TARA: It's like when Sturla proposed to Miriam. He said he wanted her to be the protagonist of his stories ...

MIRIAM: Yes, exactly! What if we could make the planet the protagonist of *our* story?

TARA: If we can love the planet enough to make it the hero, then we'll be able to change this narrative.

DAVID: Ah, I see – the planet is the hero.

STURLA: Oh! And we're just the sidekicks!

TARA: Now that's poetic.

MIRIAM: That's a story I'd want to tell my kids.

TARA: Yeah, it's a much better story than Dr. Doom and Gloom over here, saying 'It's too late' and we're all going to be dead ...

DAVID: You know, a reporter once asked me what I hope people will say about me after I'm gone. And I said, 'You mean after I am dead?'

And he said, 'Well, yeah.'

And I told him, 'I don't give a shit what people say, I'll be dead!'

What I hope is that, at the end of my life, I'm not in pain, and my grandchildren can gather around me so I can look them each in the eye and say, 'I'm only one person. I love you. And I did the best I could.'

(*Pause.*)

TARA: Well, after you die, I'll probably still be around, so I'll just ... I guess I'll finish the job.

DAVID: You'd better!

INTERLUDE
'HOPE' IS THE THING WITH FEATHERS

'Hope' is the thing with feathers –
That perches in the soul –
And sings the tune without the words –
And never stops – at all –

And sweetest – in the Gale – is heard –
And sore must be the storm –
That could abash the little Bird
That kept so many warm –

I've heard it in the chillest land –
And on the strangest Sea –
Yet – never – in Extremity,
It asked a crumb – of me.

Emily Dickinson

AFTERWORD

by Tara Cullis

One thing about David is that he's always up for everything.

Over the past forty-plus years that he's been doing *The Nature of Things*, people have asked him to do remarkable things. Like hanging off the strut of an airplane in flight or jumping into the midwinter ocean for a show on hypothermia. He thought he was being professional; I thought he was crazy. I'm a much more cautious person.

Similarly, when a stranger suggests making a feature film with him or televising him on trial for 'seditious libel,' David's happy to entertain the idea, while I envision what can possibly go wrong.

I'm much more private. too. I don't mind David going out onstage, but I'm not used to it for myself. When the Why Not team approached us about this play, I was nervous about being in the public eye.

But I've finally realized that people have brilliant ideas. David has benefited from listening to these ideas. They've helped him build a much wider audience. And plays and the theatre provide a whole new medium. They offer a different way to get our message out, to reach a brand-new audience. We knew from decades of experience that just telling people the problems and suggesting solutions doesn't work – yet we kept right on doing that over and over! So then I thought, 'I can be brave! Let's see what playwrights and actors can do.'

Writers, actors, musicians, and artists have always been at the forefront of revolutions, and a revolution is what we need right now. We have to radically change our ways. And if we're going to succeed in decarbonizing – getting off fossil fuels and staying under 1.5 °C of global warming – we need to use every tool in the kit. As it turns out, putting together a play has opened up a whole new world to us, a world of people with new tools, tricks, talent, and brilliance.

Ironically, my academic life was spent in Comparative Literature. I've read pretty near every play ever written in English, French, and German, and most of the novels and poetry too. Moreover, in my thesis I'd analyzed how lopsided life in the twentieth and twenty-first centuries has become: our romance with technology and money and individual freedom – with exploitation, extraction, and consumption – has disempowered writing, the arts, community, and the feminine, along with respect, connection, nurturing. But I'd left that immersion in literature and ideas behind when I joined the environmental movement.

It was sweet to bring the two worlds together. It made my life make more sense. And, of course, environmental work is all about respect, connection, nurturing. The feminine. Responsibility. Community.

... Love?

David and I have spent our working lives trying to communicate environmental problems. Our big concern has always been how to get the public on side. We've constantly asked ourselves, 'How can we communicate science in a way that changes public behaviour?' The surprising answer that emerged from our conversations in creating this play was ***Love***. Love for our children, love for the planet that created us, love for each other. Love is the opposite of greed and destruction and irresponsibility. Love brings us together, protecting us all, and it creates life and hope. It is immensely strong and extraordinarily powerful. Once we commit to doing what it takes to protect and nurture each other and the nature that keeps us alive, love will change the world.

STURLA ALVSVAAG

Performer

Sturla Alvsvaag is an actor and theatre-maker from Bergen, Norway. A graduate of École internationale de théâtre Jacques Lecoq in Paris, Sturla has created and performed with companies from across Australia, France, Norway and Canada. Select works include: *Traversée de la Riviére* (Collectif 2222, France), *Il Turco In Italia* (Bergen National Opera, directed by Mark Lamos), and *Wendy and Peter Pan* (the National Theatre of Norway). Sturla is also the co-artistic director of YVA Theatre Company with whom he has acted, co-created, and produced *The Nose* (commissioned by the MiniMidiMaxi Festival) and *The First Time I Saw the Sea* (Frontlosje-festivalen). In Canada, Sturla is working with Why Not Theatre and 1S1 Collective on a new adaptation of *Lady Macbeth.*

DR. TARA CULLIS

Co-Writer, Performer

An award-winning writer and former faculty member of Harvard University, Tara Cullis has been a key player in environmental movements in the Amazon, Southeast Asia, Japan, and British Columbia.

In 1990 Dr. Tara Cullis co-founded, with Dr. David Suzuki, the David Suzuki Foundation to 'collaborate with Canadians from all walks of life including government and business, to conserve our environment and find solutions that will create a sustainable Canada through science-based research, education, and policy work.'

Tara founded or co-founded nine other organizations before co-founding the David Suzuki Foundation.

She was a founder of the Turning Point Initiative, now known as the Coastal First Nations Great Bear Initiative, which brought First Nations of British Columbia's central and northern coasts into a historic alliance, protecting the ecology of the region known as the Great Bear Rainforest. Tara has been adopted and named by the Haida, Gitga'at, Heiltsuk, and Nam'gis First Nations.

MIRIAM FERNANDES

Associate Director, Co-Writer, Performer

Miriam is the co-artistic director of Why Not Theatre and has worked as an actor, director, and theatre-maker around the world. Recent directing and creation credits include *Metamorphoses* (CDTPS, University of Toronto), *Hayavadana* (Soulpepper Theatre), *Nesen* (MiniMidiMaxi Festival, Norway), and *The First Time I Saw the Sea* (YVA Company, Norway). She is currently co-writing/adapting for the stage the ancient epic *Mahabharata* (Why Not Theatre/Shaw Festival), is developing a Deaf/hearing production of *Lady Macbeth* (in partnership with 1S1 Collective), and working on a new adaptation of *The Visit*. Miriam is the recipient of the JBC Watkins Award and was nominated for the inaugural Johanna Metcalf Performing Arts Prize. Miriam has trained with Anne Bogart's SITI Company, and is a graduate of École internationale de théâtre Jacques Lecoq in Paris.

RAVI JAIN

Director, Co-writer, Original
Performance Concept

Toronto-based stage director Ravi Jain is a multi-award-winning artist known for making politically bold and accessible theatrical experiences in both small indie productions and large theatres. As the founding artistic director of Why Not Theatre, Ravi has established himself as an artistic leader for his inventive productions, international producing/collaborations, and innovative producing models, which are aimed to better support emerging artists to make money from their art.

Ravi was twice shortlisted for the 2016 and 2019 Siminovitch Prize, and won the 2012 Pauline McGibbon Award for Emerging Director and the 2016 Canada Council John Hirsch Prize for direction. He is a graduate of the two-year program at École internationale de théâtre Jacques Lecoq. He was selected to be on the roster of clowns for Cirque du Soliel.

Upcoming: Ravi's bilingual English/ASL production of *Hamlet* will tour through the U.S. with a final stop at Robert Lepage's Le Diamant in Quebec City, and in Spring 2023 he will be directing his co-adaptation of the ancient Indian epic *Mahabharata*.

DR. DAVID SUZUKI

Co-writer, Performer, Award-Winning Scientist, Environmentalist, and Broadcaster

David Suzuki is a scientist and Emeritus Professor of Genetics at the University of British Columbia. Through radio, television (*The Nature of Things*), and fifty-five books, he has communicated humanity's collective impact on the natural world, an impact that now threatens the future of human life. He is a Companion of the Order of Canada, and has received UNESCO's Kalinga Prize for Science, the Right Livelihood Award for 2009, the Global 500, and thirty honorary degrees from Canada, the U.S., and Australia. Dr. Suzuki has been honoured with adoption and names from eight Indigenous First Nations in Canada and Australia.

KEVIN MATTHEW WONG

Associate Director, Dramaturg, Producer

Kevin Matthew Wong is a theatre creator, performer, producer, facilitator, and environmentalist. He is the co-founder and Artistic Director of Broadleaf Theatre, which brings climate-justice considerations to live performance and offers dramaturgy and consultation to climate-focused artistic projects. His work *The Chemical Valley Project* is a multimedia solo performance about environmentalism and reconciliation, created in collaboration with Aamjiwnaang First Nation Water Protectors Vanessa Gray and Beze Gray, and co-creator Julia Howman. Kevin has worked with Cahoots Theatre, the Koffler Centre, Music Picnic, Theatre Passe Muraille, and the social justice residency the Gardarev Center, among others. Kevin currently facilitates the Festival Creative Producers and Administrators program at the Paprika Festival and is the Managing Producer of Why Not Theatre's Make Platform. kevinmatthewwong.com

LAND ACKNOWLEDGEMENT

This project was filmed on Quadra Island, the territory of the Laich-kwil-tach: the We Wai Kai, Wei Wai Kum, and the Kwai-kah Nations.

Why Not Theatre is based in Tkaronto, Dish With One Spoon Treaty Territory. Tkaronto is the land of the Wendat, Anishinaabe, Haudenosaunee, and Mississaugas of the Credit.

We acknowledge all of the storytellers, knowledge keepers, and caretakers who have stewarded this land from time immemorial and will continue to do so far into the future.

Typeset in Albertan and Greycliff.

Printed at the Coach House on bpNichol Lane in Toronto, Ontario, on EarthChoice cream paper. This book was printed with vegetable-based ink on a 1973 Heidelberg KORD offset litho press. Its pages were folded on a Baumfolder, gathered by hand, bound on a Sulby Auto-Minabinda, and trimmed on a Polar single-knife cutter.

Coach House is on the traditional territory of many nations, including the Mississaugas of the Credit, the Anishnabeg, the Chippewa, the Haudenosaunee, and the Wendat peoples, and is now home to many diverse First Nations, Inuit, and Métis peoples. We acknowledge that Toronto is covered by Treaty 13 with the Mississaugas of the Credit. We are grateful to live and work on this land.

Edited by Alana Wilcox
Cover design by Crystal Sikma
Wordmark design by Abigail Fleur Aries
Interior design by Crystal Sikma

Coach House Books
80 bpNichol Lane
Toronto ON M5S 3J4
Canada

416 979 2217
800 367 6360

mail@chbooks.com
www.chbooks.com